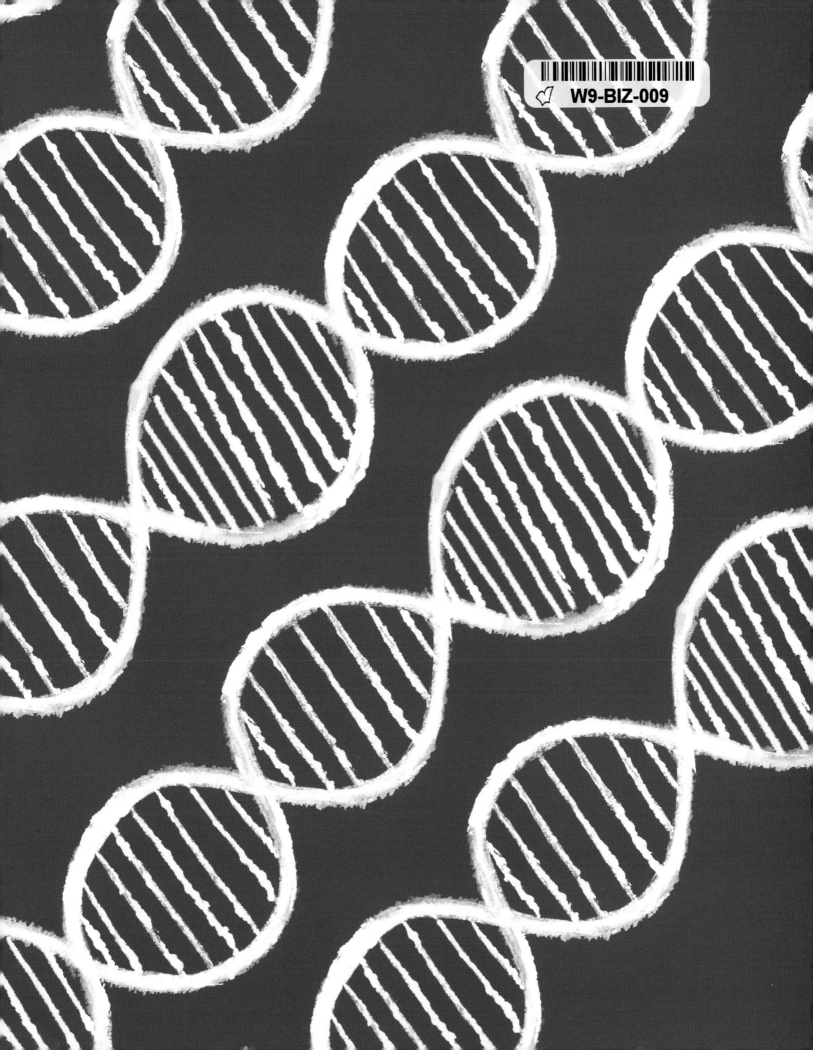

NEVER GIVE UP

DR. KATI KARIKÓ AND THE RACE FOR THE FUTURE OF VACCINES

DEBBIE DADEY

ILLUSTRATED BY JULIANA OAKLEY

MILLBROOK PRESS / MINNEAPOLIS

FOR ALEX AND FOR ALL KIDS WHO
ARE CURIOUS, HAVE A DREAM, AND
WILL NOT BE DETERRED —D.D.

TO ALL THOSE WHO FOLLOW THEIR HEART
AND TO THOSE WHO SUPPORT THEM —J.O.

Text copyright © 2023 by Debbie Dadey
Illustrations copyright © 2023 by Juliana Oakley

Millbrook Press™
An imprint of Lerner Publishing Group, Inc.
241 First Avenue North
Minneapolis, MN 55401 USA

For reading levels and more information, look up this title at www.lernerbooks.com.

Backmatter photos provided by: Dr. Katalin Karikó and her daughter, Zsuzsanna (Susan) Francia
(family photos); Penn Medicine (vaccine photo).

Designed by Lindsey Owens.
Main body text set in Breughel Com 55 Roman. Typeface provided by Linotype AG.
The illustrations in this book were created with pencil, Procreate, and Photoshop.

Library of Congress Cataloging-in-Publication Data

Names: Dadey, Debbie, author. | Oakley, Juliana, illustrator.
Title: Never give up : Dr. Kati Karikó and the race for the future of vaccines / Debbie Dadey ; illustrated by
    Juliana Oakley.
Description: Minneapolis : Millbrook Press, [2023] | Includes bibliographical references. | Audience: Ages
    5–10 | Audience: Grades 2–3 | Summary: "This picture book biography introduces Hungarian American
    biochemist Katalin Karikó, who played a critical role in developing the mRNA vaccine for COVID-19. Follow
    the journey of Katalin (Kati) Karikó from her childhood in rural Hungary" —Provided by publisher.
Identifiers: LCCN 2022020318 (print) | LCCN 2022020319 (ebook) | ISBN 9781728456331 (library
    binding) | ISBN 9781728485584 (ebook)
Subjects: LCSH: Karikó, Katalin—Juvenile literature. | Biochemists—Biography—Juvenile literature. |
    Hungarian Americans—Biography—Juvenile literature.
Classification: LCC QP511.8.K37 D33 2023 (print) | LCC QP511.8.K37 (ebook) |
    DDC 572.092 [B]—dc23/eng/20220624

LC record available at https://lccn.loc.gov/2022020318
LC ebook record available at https://lccn.loc.gov/2022020319

Manufactured in the United States of America
1-50767-50125-6/28/2022

BY THE TIME THE MORNING SUN SHONE ON THE REED ROOF OF KATI'S ONE-ROOM HOME IN HUNGARY, she had already fed the chickens, collected eggs, and been chased by a rooster. She stopped to watch the sow who was going to have babies very soon. Kati hated to miss it, but she didn't want to be late for school. With any luck, Apuka, her dad, would let her help with the piglets in the afternoon.

Apuka taught her about animals at home, but Kati learned even more at school. One day her teacher brought a strange-looking metal object into the classroom. Her teacher showed the class how to look into the microscope to see little blobs called cells. Kati couldn't believe her whole body was made up of these tiny things. In fact, *everything* living is composed of cells—even grouchy roosters!

At recess, Kati raced her friends. She was a fast runner, but she didn't win every time. She never gave up though. After recess, she was eager to get back inside to find out more about cells and the scientists who studied them.

"AS LONG AS I WAS IN THE LAB, I HAD FUN. IT WAS JUST SUCH A JOY. EVEN IF THINGS DIDN'T WORK OUT AS I EXPECTED."

—Dr. Kati Karikó

That evening, Kati told her family she wanted to be a scientist. Apuka was a butcher, and Anyuka, Kati's mother, did the bookkeeping at a nearby bottle factory. They believed that with hard work and perseverance, their children could do anything.

Anyuka said, "I look forward to you earning the Nobel Prize."
Kati loved a challenge. Could she really win the biggest honor possible for a scientist?

While she was growing up, Kati went to science camps. At one camp, they visited a fish hatchery. Kati found out that some types of food made the fish grow bigger.

Making one small change could have a huge impact.

When she was fourteen, Kati's teacher chose her and another student to go to the Science Olympics. It was a weeklong competition on the other side of the country. Kati and the younger boy would have to ride on a train all by themselves.

But there was a big problem: That same week, Kati's sister and Anyuka were going on vacation to a city with a water park. It would be so fun, but Kati couldn't do both. She chose the science competition.

On the day of the Science Olympics, the buildings got taller and taller as the train pulled closer to their stop in Budapest. It was the biggest city Kati had ever seen.

At the competition, Kati was amazed at how smart the other kids were. She came in third place, but there was still so much she needed to learn. Instead of scaring her, the experience made her excited to study more.

When she got older, Kati learned that every cell contains DNA, or deoxyribonucleic acid. This substance contains the instructions for how to build and maintain a living thing. Children inherit DNA from their parents, which is why the color of your eyes or hair might be the same as your parents or grandparents. It is also the reason you are tall or short, and why you have a little nose or big ears.

Kati got married and had a daughter whose greenish-blue eyes matched hers. By then, Kati had finished school and was known as Dr. Katalin Karikó.

At her first job, in a biology research center, Kati became curious about mRNA, or messenger ribonucleic acid. It is a hardworking copy of a part of DNA. It provides instructions to make proteins. When sickness strikes, the body responds by making proteins—such as germ-attacking antibodies—to keep us healthy.

"SHE WAS, IN A POSITIVE SENSE, KIND OF OBSESSED WITH THE CONCEPT OF MESSENGER RNA. IT'S GOING TO BE TRANSFORMING."

—Dr. Anthony Fauci, director of the National Institute of Allergy and Infectious Diseases

Kati thought that this natural process of fighting illness could be improved. After all, sometimes our bodies are overwhelmed by a disease they are not prepared to fight. If Kati could make a new mRNA that would send the right message to make the right proteins, would people get well faster or maybe not get sick at all?

Unfortunately, the body destroyed any mRNA that it hadn't created itself. Other scientists told Kati it was impossible to teach cells to cure themselves.

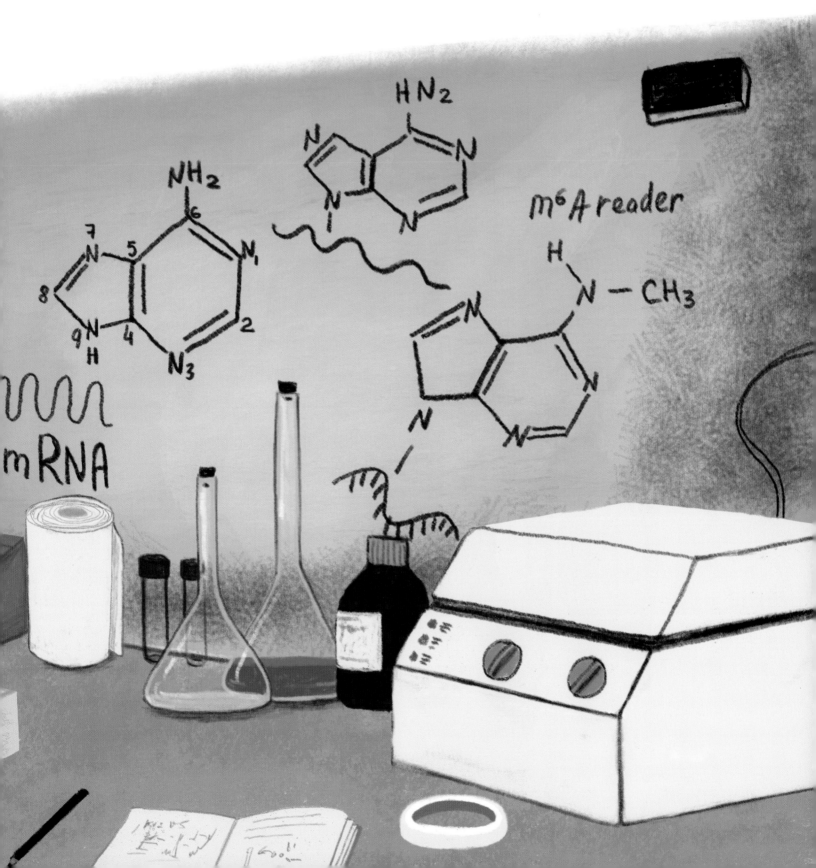

Kati needed to move to the United States, where she could learn from other scientists who studied mRNA. Her family packed for their big move—but they couldn't bring much. The Communist government in Hungary forced her to leave most of her things behind. She hid money in her two-year-old daughter's teddy bear and took her family across the ocean to begin a job at Temple University. They needed that money to begin their new life!

But at her new job, Kati heard the same thing from fellow scientists: "You'll never succeed in making medicine with mRNA." Some even laughed at her. Should she give up? Was she smart enough to figure it out?

WAS SHE WRONG TO BE CURIOUS?

Kati believed her work was important, so she spent years doing more experiments. But no matter what she tried, cells fought her mRNA instead of letting it help. It was a difficult time. One scientist wanted to see the male boss of Dr. Karikó's laboratory, not believing that a woman could be in charge. Another wanted to kick her out of the United States. She was fired from Temple University and later demoted at the University of Pennsylvania because her bosses thought her idea was a dead end.

"WHEN I AM KNOCKED DOWN, I KNOW HOW TO PICK MYSELF UP."

—Dr. Kati Karikó

Kati woke up at five o'clock every morning to study in her home office before she went to her job. She also worked on holidays and weekends. Sometimes, she'd snack on her favorite treat, Goobers chocolate-covered peanuts, as she worked. Her husband, Béla Francia, knew she loved to research. He told her, "You're not going to work—you are going to have fun." Often she stayed up late at night writing grant proposals to ask for research money. It cost a lot to do experiments and to have a laboratory full of equipment.

When she needed a break, she went running. Her now-grown daughter, Susan, would often bike beside her. Kati even ran marathons. She never quit until she'd finished the race. While she ran, she would think about how to make her mRNA idea work.

One day in 1997, Kati went to the copy machine at the university where she worked. Another scientist, Drew Weissman, wanted to use the machine too. Kati told him about her work. They talked and came up with a new idea to alter one of the four parts of mRNA so it could train the body to beat a virus or treat a disease. The idea was like changing a tire on a car but on a very small scale. Would it work? It took eight years, but . . .

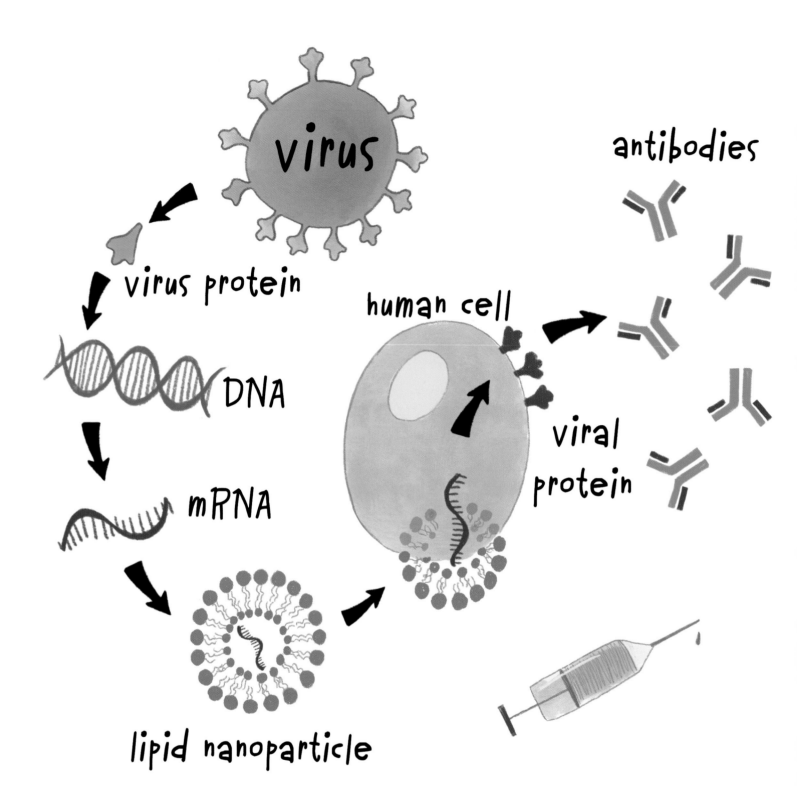

virus

antibodies

virus protein

human cell

DNA

viral
protein

mRNA

lipid nanoparticle

It worked! Kati couldn't believe it. Her experiment was a success. She tried it again and again, and it worked every time! Tiny balls of fat called lipid nanoparticles protected the mRNA and carried it into cells. Drew and Kati wrote a paper about their breakthrough. They had found a way to make mRNA that could teach cells to create a specific virus protein. That protein would prompt the body to make antibodies to fight against a virus or could be therapeutic and cure a disease.

Many other scientists were astounded at their achievement. One was so impressed that he helped to create a whole new company, called Moderna, based on their ground-breaking work.

Kati was offered jobs at BioNTech and Moderna, both companies that were experimenting with mRNA medications. She would get research funding to discover even more about mRNA and how it could prevent or treat specific diseases. Kati chose BioNTech. To work there, she would have to move to Germany for a while. She would miss her family, but she knew that her work might eventually save lives. So once again, she packed her bags and moved to a far-off country.

"I FEEL HUMBLED, AND HAPPY. I AM MORE OF A BASIC SCIENTIST, BUT I ALWAYS WANTED TO DO SOMETHING TO HELP PATIENTS."

—Dr. Kati Karikó

In 2020, a disease called COVID-19 spread all over the world. Hundreds of millions of people got sick, and many died. To protect themselves, people wore masks. Stores and restaurants closed. Few people flew on planes. Many adults did their jobs from home, and kids had school online. Researchers worldwide worked to figure out how to stop the virus with a vaccine. If a vaccine could keep people from getting sick and spreading COVID-19 to others, the pandemic would no longer be so dangerous.

CANCELED DUE TO COVID

CANCELED

In the past, vaccines had usually been made using weakened or dead viruses, but Kati believed that mRNA could get the job done quicker. Her fellow scientist, Drew Weissman, agreed. He said, "We knew when we started with this technology that it would be very useful if a pandemic hit, because it's so fast and so easy to make a vaccine with it."

Scientists at BioNTech and a large company called Pfizer worked together on Kati's idea, and so did Moderna, while other companies used traditional methods. Thanks to Kati's research, Pfizer and BioNTech were able to design a vaccine within hours after learning the DNA code of the COVID-19 virus. Moderna made an mRNA vaccine as well. But both vaccines still had to be tested. Because it was an emergency, many tests were done at the same time. What could have taken years was finished in ten months. Everyone wanted to know if the mRNA vaccines would work.

THEY DID! Tests showed that the Pfizer-BioNTech vaccine was extremely effective. Kati was so excited by the results that she celebrated by eating a whole box of Goobers.

The vaccine reduces a person's chance of getting sick, and it lessons an infected person's chance of becoming seriously ill.

"I'M HOPEFUL, NOW THAT THERE IS SO MUCH INTEREST AND EXCITEMENT FOR THIS RESEARCH, THAT IT WILL BE POSSIBLE TO DEVELOP AND TEST THIS mRNA VACCINE TECHNOLOGY FOR PREVENTION AND TREATMENT OF OTHER DISEASES, TOO."

—Dr. Kati Karikó

Curious Kati helped the world recover from a terrible pandemic all because she wouldn't quit. Her forty years of hard work have finally paid off, but Kati isn't finished with her race. She hopes to find out if mRNA can cure or prevent other illnesses and believes one day we might keep some in our homes to use as treatment when we get sick.
SHE IS STILL CURIOUS!

TIMELINE

1982: Kati gives birth to a daughter, Susan Francia.

Kati finishes her PhD at the University of Szeged in Hungary. She continues her research in the Institute of Biophysics at the Biological Research Centre of Hungary.

1988: Kati loses her job at Temple University.

1989: Kati begins work at the University of Pennsylvania.

1955: Biochemist Dr. Katalin Karikó is born in Szolnok, Hungary, on January 17 and grew up in the tiny town of Kisújszállás.

1961: Scientists first discover mRNA.

1978: Kati first focuses on mRNA at the Szeged Biological Research Centre in Hungary.

1980: Kati marries Béla Francia on October 11.

1985: Kati leaves Hungary with money hidden in her daughter's teddy bear to be a postdoctoral fellow at Temple University in Philadelphia.

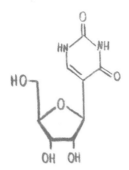

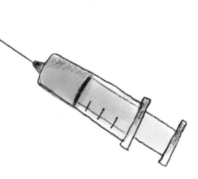

1990: Kati submits her first mRNA grant proposal.

1995: Kati is demoted from faculty position of assistant professor to research investigator.

2006–2013: Kati cofounds and becomes the CEO of RNARx, a biotech company.

2008 AND 2012: Kati's daughter, Susan Francia, wins Olympic gold as part of the US National Rowing Team.

2010: Moderna is founded.

DECEMBER 2020: A ninety-year-old British woman receives the first Pfizer/BioNTech COVID-19 vaccine as part of a mass vaccination program. Kati and Drew receive their first dose of the Pfizer/BioNTech on December 18, just days after it was approved for emergency use in the United States for people ages sixteen and older.

2021: The Pfizer/BioNTech vaccine is approved for ages twelve to fifteen in May. It is approved for ages five to eleven in November.

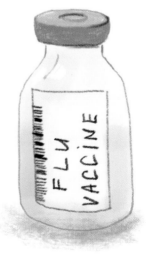

1997: Kati begins working with Dr. Drew Weissman.

2005: Kati and Drew publish a study that opens the door for mRNA to be used in vaccines.

2013: Kati retires from the University of Pennsylvania and becomes a senior vice president at BioNTech.

2018: BioNTech partners with Pfizer to develop an mRNA-based flu vaccine.

2020: BioNTech partners with Pfizer to make a COVID-19 vaccine, using the mRNA technology.

STEPS TO MAKING A VACCINE

Developing a vaccine can take ten to fifteen years—or longer. After that, it still must be tested and approved before it can be given to people. Luckily, Katalin Karikó had already done much of the exploratory stage when the need for a COVID-19 vaccine arose.

An mRNA vaccine can be mass-produced much faster than traditional ones such as the polio vaccine, which takes eighteen months to make. That's because with mRNA vaccines, an actual virus is not used, only instructions to prevent the infection. Pfizer/BioNTech can make three million doses in just sixty days.

STAGES ALL VACCINES IN THE UNITED STATES MUST GO THROUGH BEFORE THEY ARE GIVEN TO THE PUBLIC

1. EXPLORATORY STAGE:

An idea is researched and developed.

2. PRECLINICAL STAGE:

The first experiments take place, and vaccines are given to animals.

3. CLINICAL STAGE:

Vaccines are given to human volunteers after the Food and Drug Administration (FDA) reviews and approves. If the FDA does not approve at any point, the process is stopped or delayed until changes are made and approval is received.

Phase 1: Small groups are given the vaccine, and the results are studied. The FDA reviews and approves the next step.

Phase 2 and 3: Larger groups of certain ages and health conditions are given the vaccine, and the results are studied. The FDA reviews and approves the next step.

Phase 4: Thousands of people are given the vaccine, and the results are studied.

4. REGULATORY REVIEW AND APPROVAL:

All clinical results are presented to the FDA to make sure the vaccine is safe. The FDA reviews the results of all the experiments. They make sure the place where it is to be made is clean and safe before granting approval.

5. MANUFACTURING AND QUALITY CONTROL:

Drug manufacturing sites are continually monitored and checked for cleanliness and proof that the drug is being made correctly. This monitoring begins even before the phase 1 trial.

AUTHOR'S NOTE

Dr. Katalin Karikó's pioneering work with mRNA inspired me, and I wanted other people to know what she achieved. She endured great hardships as a result of her unending faith that mRNA would prove invaluable to helping people. She found many things that wouldn't work, but after never giving up on mRNA, she found the thing that did. Her curiosity and tenacity saved many lives during the COVID-19 pandemic.

For a lot of people, the years 2020 and 2021 were terrible. It was for my family. We lost my beloved youngest son, Alex Dadey, as a result of the sad times. Many families, like mine, will never recover from their losses, but I am thankful to scientists like Kati who want to make the world better and don't give up even when times are hard.

GLOSSARY

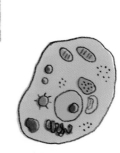

ANTIBODY: a protein that binds and inactivates (neutralizes) harmful bacteria and viruses

CELL: the smallest unit that has all the basic properties of life. Some tiny creatures are made up of just one cell. You have 210 different kinds of cells in your body, including blood cells, bone cells, and cells that make muscles.

COVID-19: a new illness caused by a virus, first infecting humans in late 2019. It has spread throughout the world, making it the worst pandemic since the influenza pandemic of 1918.

DNA (DEOXYRIBONUCLEIC ACID): a part of the cell that contains all the information an organism needs to develop and to survive. DNA is passed from parent to child.

LIPID NANOPARTICLE: a very tiny ball of fat. The BioNTech/Pfizer and Moderna COVID-19 vaccines both use lipid nanoparticles to protect mRNA and carry it into cells.

MRNA (MESSENGER RIBONUCLEIC ACID): a single-stranded molecule that carries DNA messages to the protein-making part of each cell

PANDEMIC: a disease that spreads across multiple countries and continents, usually without a known cure

PROTEIN: a large molecule made up of amino acids. Proteins are a hardworking part of cells. Some proteins are antibodies that help protect the body from viruses and bacteria.

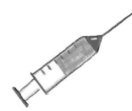

VACCINE: a substance that is put into a body, usually through a shot, that helps the body fight off a germ and prevent infection

VIRUS: a microbe that can make people very sick. Viruses can survive and reproduce only inside cells.

SOURCE NOTES

"As long as . . . I expected": Vicky Feng, "How Covid Vaccine Tech Could Soon Be Used to Fight Cancer," Bloomberg, April 9, 2021, https://www.bloomberg.com/news/videos/2021-04-09/how-covid-vaccine-tech-could-soon-be-used-to-fight-cancer-video.

"The best scientists . . . to ask": Gina Kolata, "Kati Kariko Helped Shield the World from the Coronavirus," *New York Times*, last modified April 17, 2021, https://www.nytimes.com/2021/04/08/health/coronavirus-mrna-kariko.html.

"She was . . . transforming": Kolata.

"When I am . . . pick myself up": Sarah Newey and Paul Nuki, "Redemption': How a Scientist's Unwavering Belief in mRNA Gave the World a Covid-19 Vaccine," *Telegraph* (London), December 2, 2020, https://www.telegraph.co.uk/global-health/science-and-disease/redemption-one-scientists-unwavering-belief-mrna-gave-world/.

"You're not going . . . have fun": Kolata, "Kati Kariko Helped Shield the World."

"It's not a . . . passion": "mRNA Day 2020 Celebrating the Past, Present, and Future of mRNA," YouTube video, 1:20:27, posted by TriLink Technologies, December 2, 2020, https://www.youtube.com/watch?v=Eysud56Va20.

"I feel humbled . . . help patients": Penn Medicine News, "University of Pennsylvania mRNA Biology Pioneers Receive COVID-19 Vaccine Enabled by their Foundational Research," news release, December 23, 2020, https://www.pennmedicine.org/news/news-releases/2020/december/penn-mrna-biology-pioneers-receive-covid19-vaccine-enabled-by-their-foundational-research.

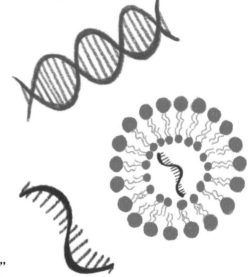

"We knew when . . . vaccine with it": Matthew DeGeorge, "The Vaccine Trenches," *Pennsylvania Gazette*, April 20, 2021, https://thepenngazette.com/the-vaccine-trenches/.

"I just . . . bear fruit": Dana Kennedy, "This Scientist's Decades of mRNA Research Led to Both COVID-19 Vaccines," *New York Post*, December 5, 2020, https://nypost.com/2020/12/05/this-scientists-decades-of-mrna-research-led-to-covid-vaccines/.

"I'm hopeful, now . . . diseases, too": Penn Medicine News, "University of Pennsylvania mRNA Biology Pioneers."

"Katalin Karikó deserves the Nobel Prize": "Katalin Karikó, the Hungarian Biochemist behind the COVID Vaccines," Kafkadesk, December 7, 2020, https://kafkadesk.org/2020/12/07/katalin-kariko-the-hungarian-biochemist-behind-the-covid-vaccines/; Kennedy, "This Scientist's Decades of mRNA Research."

FURTHER STUDY

"Biology for Kids DNA and Genes." Ducksters, accessed June 23, 2022, https://www.ducksters.com/science/biology/dna.php.

Cuomo, Chris. "She Is One of the Scientists Who Paved the Way for the COVID-19 Vaccine." CNN video, 3:57, December 15, 2020. https://www.cnn.com/videos/health/2020/12/15/katalin-karik-biontech-senior-vice-president-mrna-cpt-vpx.cnn.

"How Do COVID-19 mRNA Vaccines Work?" WebMD video, 3:09, February 9, 2021. https://www.webmd.com/vaccines/covid-19-vaccine/video/video-covid-mrna-vaccine.

"How mRNA Vaccines Work." YouTube video, 4:25, Posted by Simply Explained, December 30, 2020. https://www.youtube.com/watch?v=WOvvyqJ-vwo.

"How Vaccines Work against COVID-19: Science, Simplified." YouTube video, 2:15. Posted by Scripps Research, December 18, 2020. https://www.youtube.com/watch?v=uWGTciX795o.

Levine, Sara. *Germs Up Close*. Minneapolis: Millbrook Press, 2021.

Marshall, Linda Elovitz. *The Polio Pioneer*. New York: Knopf Books for Young Readers, 2020.

Messner, Kate. *Dr. Fauci: How a Boy from Brooklyn Became America's Doctor*. New York: Simon and Schuster Books for Young Readers, 2021.

Pham, LeUyen. *Outside, Inside*. New York: Roaring Book, 2021.

Slade, Suzanne. *June Almeida, Virus Detective! The Woman Who Discovered the First Human Coronavirus*. Ann Arbor, MI: Sleeping Bear, 2021.

"What Is DNA for Kids/An Easy Overview of DNA for Children/Awesome DNA Facts." YouTube video, 5:25. Posted by Learn Bright, March 6, 2019. https://www.youtube.com/watch?v=921XdtoRAoo.

Woollard, Alison, and Sophie Gilbert. *The DNA Book*. London: Dorling Kindersley, 2020.

ACKNOWLEDGMENTS

Thanks to my immunologist daughter, Dr. Rebekah Dadey; my pharmaceutical scientist husband, Dr. Eric Dadey; and consultant Dr. Kaitlyn Morabito for their knowledge and insights.

Grateful thanks to Dr. Katalin Karikó and her daughter, Zsuzsanna (Susan) Francia, for sharing their stories and expert advice. Congratulations to Dr. Karikó for being awarded many prizes, including the Wilhelm Exner Medal from the Austrian Trade Association and the Széchenyi Prize by the Hungarian government. A cofounder of Moderna, Derrick Rossi, said, "Katalin Karikó deserves the Nobel Prize."

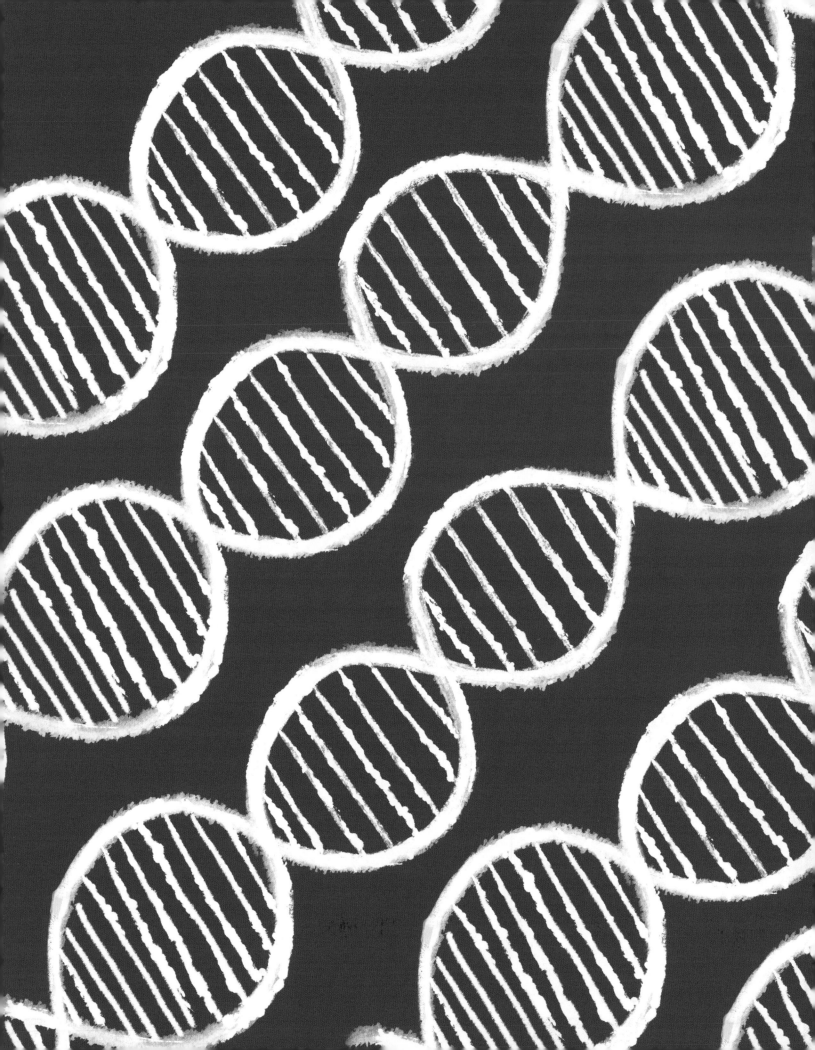